Weekly Schedule

Time / Period	Monday	Tuesday	Wednesday	Thursday	Friday	Saturday	Sunday

AF887862

Emergency Contacts

Name:	Relation:	Cell Phone:	Phone:	Email:
Address:		City:	State:	Zip Code:

Name:	Relation:	Cell Phone:	Phone:	Email:
Address:		City:	State:	Zip Code:

Academic Calendar

Year:

SPRING SEMESTER 2

Day														

January

1	2	3	4	5	6	7	8	9	10	11	12	13	14
15	16	17	18	19	20	21	22	23	24	25	26	27	28
29	30	31											

February

			1	2	3	4	5	6	7	8	9	10	11
12	13	14	15	16	17	18	19	20	21	22	23	24	25
26	27	28	29										

Change day if it is a leap year (2024, 2028, 2032, 2036, 2040, 2044, 2048, 2052, 2056, 2060. . .)

Day														

March

			1	2	3	4	5	6	7	8	9	10	11
12	13	14	15	16	17	18	19	20	21	22	23	24	25
26	27	28	29	30	31								

April

						1	2	3	4	5	6	7	8
9	10	11	12	13	14	15	16	17	18	19	20	21	22
23	24	25	26	27	28	29	30						

SUMMER SEMESTER 3

May

	1	2	3	4	5	6	7	8	9	10	11	12	13
14	15	16	17	18	19	20	21	22	23	24	25	26	27
28	29	30	31										

June

				1	2	3	4	5	6	7	8	9	10
11	12	13	14	15	16	17	18	19	20	21	22	23	24
25	26	27	28	29	30								

July

						1	2	3	4	5	6	7	8
9	10	11	12	13	14	15	16	17	18	19	20	21	22
23	24	25	26	27	28	29	30	31					

FALL SEMESTER 1

August

		1	2	3	4	5	6	7	8	9	10	11	12
13	14	15	16	17	18	19	20	21	22	23	24	25	26
27	28	29	30	31									

September

					1	2	3	4	5	6	7	8	9
10	11	12	13	14	15	16	17	18	19	20	21	22	23
24	25	26	27	28	29	30							

October

1	2	3	4	5	6	7	8	9	10	11	12	13	14
15	16	17	18	19	20	21	22	23	24	25	26	27	28
29	30	31											

November

			1	2	3	4	5	6	7	8	9	10	11
12	13	14	15	16	17	18	19	20	21	22	23	24	25
26	27	28	29	30									

December

					1	2	3	4	5	6	7	8	9
10	11	12	13	14	15	16	17	18	19	20	21	22	23
24	25	26	27	28	29	30	31						

Events and Due Dates

Title:	Start:	End:
Note:		

Title:	Start:	End:
Note:		

Title:	Start:	End:
Note:		

Title:	Start:	End:
Note:		

Title:	Start:	End:
Note:		

Title:	Start:	End:
Note:		

Assignment / Project / Exam	Due Date

Assignment / Project / Exam	Due Date

Copyright & Disclaimer

NOTABLE+ CHEMISTRY NOTEBOOK: G3E, COOL DRIP, STANDARD+E (120 PAGES)

Published by Notable Plus LLC.

The author is represented by Notable Plus LLC, Palm Bay, Florida.

Visit our website at www.notableplus.com

Notable+ Chemistry Notebook: G3E, Cool Drip, Standard+E (120 Pages)
Paperback: 120 Pages
Print Quality: Black ink and 55# (90 GSM) white paper
ISBN-10: 1-960176-12-9
ISBN-13: 978-1-960176-12-7

Notable+ also publishes it books and content in a variety of electronic formats and online. Some of the content that appears in print may not be available in electronic books or online and some of the content that appears online or in electronic formats may not be available in print.

Contents (Fillable)

Notables Contents

My Notables

Date: / /

Date: / /

Date: / /

Date: / /

Date: / /

Date: / /

Date: / /

Date: / /

Date: / /

Date: / /

Date: / /

Date: / /

Date: / /

Date: / /

Date: / /

Date: / /

Date: / /

Date: / /

Date: / /

Date: / /

Date: / /

Date: / /

Date: / /

Date: / /

Date: / /

Date: / /

Date: / /

Date: / /

Date: / /

Date: / /

Date: / /

Date: / /

Date: / /

Date: / /

Date: / /

Date: / /

Date: / /

Date: / /

Date: / /

Date: / /

Date: / /

Date: / /

Date: / /

Date: / /

Date: / /

Date: / /

Date: / /

Date: / /

Date: / /

Date: / /

Date: / /

Date: / /

Date: / /

Date: / /

Date: / /

Date: / /

Date: / /

Date: / /

Date: / /

Date: / /

Date: / /

Date: / /

Date: / /

Date: / /

Date: / /

Date: / /

Date: / /

Date: / /

Date: / /

Date: / /

Date: / /

Date: / /

Date: / /

Date: / /

Chemistry Notables

1 Periodic Table

A table of the chemical elements arranged by atomic number.

Actinides	Alkali metals	Alkaline earth metals	Lanthanides	Metalloids
Noble gases	Post-transition metals	Reactive nonmetals	Transition metals	Unknown properties

1 H																	2 He
3 Li	4 Be											5 B	6 C	7 N	8 O	9 F	10 Ne
11 Na	12 Mg											13 Al	14 Si	15 P	16 S	17 Cl	18 Ar
19 K	20 Ca	21 Sc	22 Ti	23 V	24 Cr	25 Mn	26 Fe	27 Co	28 Ni	29 Cu	30 Zn	31 Ga	32 Ge	33 As	34 Se	35 Br	36 Kr
37 Rb	38 Sr	39 Y	40 Zr	41 Nb	42 Mo	43 Tc	44 Ru	45 Rh	46 Pd	47 Ag	48 Cd	49 In	50 Sn	51 Sb	52 Te	53 I	54 Xe
55 Cs	56 Ba	57 La	72 Hf	73 Ta	74 W	75 Re	76 Os	77 Ir	78 Pt	79 Au	80 Hg	81 Tl	82 Pb	83 Bi	84 Po	85 At	86 Rn
87 Fr	88 Ra	89 Ac	104 Rf	105 Db	106 Sg	107 Bh	108 Hs	109 Mt	110 Ds	111 Rg	112 Cn	113 Nh	114 Fl	115 Mc	116 Lv	117 Ts	118 Og

58 Ce	59 Pr	60 Nd	61 Pm	62 Sm	63 Eu	64 Gd	65 Tb	66 Dy	67 Ho	68 Er	69 Tm	70 Yb	71 Lu
90 Th	91 Pa	92 U	93 Np	94 Pu	95 Am	96 Cm	97 Bk	98 Cf	99 Es	100 Fm	101 Mn	102 No	103 Lr

2 Element List

1. Hydrogen
2. Helium
3. Lithium
4. Beryllium
5. Boron
6. Carbon
7. Nitrogen
8. Oxygen
9. Fluorine
10. Neon
11. Sodium
12. Magnesium
13. Aluminium
14. Silicon
15. Phosphorus
16. Sulfur
17. Chlorine
18. Argon
19. Potassium
20. Calcium
21. Scandium
22. Titanium
23. Vanadium
24. Chromium
25. Manganese
26. Iron
27. Cobalt
28. Nickel
29. Copper
30. Zinc
31. Gallium
32. Germanium
33. Arsenic
34. Selenium
35. Bromine
36. Krypton
37. Rubidium
38. Strontium
39. Yttrium
40. Zirconium
41. Niobium
42. Molybdenum
43. Technetium
44. Ruthenium
45. Rhodium
46. Palladium
47. Silver
48. Cadmium
49. Indium
50. Tin
51. Antimony
52. Tellurium
53. Iodine
54. Xenon
55. Caesium
56. Barium
57. Lanthanum
58. Cerium
59. Praseodymium
60. Neodymium
61. Promethium
62. Samarium
63. Europium
64. Gadolinium
65. Terbium
66. Dysprosium
67. Holmium
68. Erbium
69. Thulium
70. Ytterbium
71. Lutetium
72. Hafnium
73. Tantalum
74. Tungsten
75. Rhenium
76. Osmium
77. Iridium
78. Platinum
79. Gold
80. Mercury
81. Thallium
82. Lead
83. Bismuth
84. Polonium
85. Astatine
86. Radon
87. Francium
88. Radium
89. Actinium
90. Thorium
91. Protactinium
92. Uranium
93. Neptunium
94. Plutonium
95. Americium
96. Curium
97. Berkelium
98. Californium
99. Einsteinium
100. Fermium
101. Mendelevium
102. Nobelium
103. Lawrencium
104. Rutherfordium
105. Dubnium
106. Seaborgium
107. Bohrium
108. Hassium
109. Meitnerium
110. Darmstadtium
111. Roentgenium
112. Copernicium
113. Nihonium
114. Flerovium
115. Moscovium
116. Livermorium
117. Tennessine
118. Oganesson

Chemistry Notables Level 1

3 **Intro to Chemistry** **Science**- a branch of knowledge or study dealing with a body of facts or truths systematically arranged and showing the operation of general laws.
Chemistry- the science that deals with the composition and properties of substances and various elementary forms of matter.
Chemical- a substance produced by or used in a chemical process.

4 **Atom** The smallest unit of matter. (Level 2) The smallest component of an element having the chemical properties of the element, consisting of a nucleus containing combinations of neutrons and protons and one or more electrons bound to the nucleus by electrical attraction; the number of protons determines the identity of the element.

5 **Atomic Number** The number of positive charges or protons in the nucleus of an atom of a given element.

6 **Bond** An attraction between atoms that forms chemical substances.

7 **Compound** A pure substance composed of two or more elements whose composition is constant.

8 **Electron** A negatively charged particle that orbits the nucleus of an atom.

9 **Element** One of a class of substances that cannot be separated into simpler substances by chemical means.

10 **Ion** An electrically charged atom or group of atoms formed by the loss or gain of one or more electrons.

11 **Isotope** Any of two or more forms of a chemical element, having the same number of protons in the nucleus, or the same atomic number, but having different numbers of neutrons in the nucleus, or different atomic weights.

12 **Mass Number** The integer nearest in value to the atomic weight of an atom and equal to the number of nucleons in the nucleus of the atom.

13 **Matter** The substance of which a physical object is composed.

14 **Metal** A substance that is a good conductor of heat and electricity. They are generally malleable, ductile, and shiny.

15 **Mixture** A combination of two or more substances that are not chemically combined.

16 **Molecule** The smallest physical unit of an element or compound, consisting of one or more like atoms in an element and two or more different atoms in a compound.

17 **Neutron** An elementary particle having no charge, mass slightly greater than that of a proton.

18 **Solution** A homogeneous, molecular mixture of two or more substances. (Level 2) the process by which a gas, liquid, or solid is dispersed homogeneously in a gas, liquid, or solid without chemical change.

19 **Solvent** A substance that dissolves another to form a solution.

Chemistry Notables Level 2

20 **Acid** A compound having a pH value of less than 7, usually having a sour taste and capable of neutralizing alkalis and reddening blue litmus paper.

21 **Alkali metals** Any of the group of univalent metals including potassium, sodium, lithium, rubidium, cesium, and francium, whose hydroxides are alkalis.

22 **Alkaline earth metals** The group of bivalent metals including barium, radium, strontium, calcium, and, usually, magnesium, the hydroxides of which are alkalis but less soluble than those of the alkali metals.

23 **Allotropes** One of two or more existing forms of an element: Graphite and diamond are allotropes of carbon.

24 **Alloy** A substance composed of two or more metals, or of metals with a nonmetal, intimately mixed, as by fusion or electrodeposition.

25 **Base** Is the opposite of acid and has a pH of 7 to 14. Compounds that form hydroxyl ions (OH) when dissolved in water, and whose aqueous solutions have a bitter taste and react with acids to form salts. Bases turn red litmus paper blue.

26 **Chemical reaction** A process where a set of substances undergo a chemical change to form a different substance.

27 **Covalent bond** Occurs between atoms that share a pair of electrons in order to gain full outer shells.

28 **Density** Density is the mass per unit volume of a substance.

29 **Distillation** The process of separating mixtures by boiling a liquid and then recovering the cooled gas.

30 **Electrolysis** A process used to drive a chemical reaction by using an electrical current. It is often used to separate elements.

31 **Noble gases** A family of elements that makes up column 18 of the periodic table. They are very stable due to their full outer electron shells.

32 **Oxidation** The name for the process when oxygen combines with another element to form a compound.

33 **PH scale** Measures the activity of the hydrogen ion in substances. If a substance has a pH less than 7 it is said to be acidic. If its PH is greater than 7 then it said to be basic or alkaline.

34 **Radioactivity** Or radioactive decay is a process in which an unstable nucleus loses energy by the emission of radiation in the form of a particle.

35 **Reaction** The process that occurs when two or more chemical substances are combined to form another chemical substance.

36 **Salts** Chemical compounds that are formed by the reaction of an acid and a base.

Notables (Fillable)

Ref: Page #: Line #:

Ref: Page #: Line #:

Ref: Page #: Line #:

Ref: Page #: Line #:

Notables (Fillable)

Ref: Page #: Line #:

Ref: Page #: Line #:

Ref: Page #: Line #:

Ref: Page #: Line #:

Notables (Fillable)

Ref: Page #: Line #:

Ref: Page #: Line #:

Ref: Page #: Line #:

Ref: Page #: Line #:

Notables (Fillable)

Ref: Page #: Line #:

Ref: Page #: Line #:

Ref: Page #: Line #:

Ref: Page #: Line #:

Index (Fillable)

Keyword, page#:ID#:line#

Index (Fillable)

Made in the USA
Monee, IL
08 November 2026

67946271R00070